Table of Contents

Student Name: ______________________ Notebook Number: ______________________

Email: ______________________ Phone: ______________________

Network ID: ______________________ Course: ______________________

Lab Instructor: ______________________ Section: __________ Semester: __________

Lab Partners: ____

Date	Experiment/Subject	Page Number

Table of Contents

Date	Experiment/Subject	Page Number

Laboratory Safety

Laboratory safety should be the number one priority for you, your fellow students, and your instructors. The guidelines seen below are general laboratory safety practices that can apply to most laboratory settings. Your instructor will advise you of any specific precautions or hazards you may deal with in a particular experiment. Do not hesitate to ask your instructor about any questions you might have about an experiment or any of the safety precautions outlined here or given to you by your instructor.

- Never work alone in the lab without supervision and work only on an experiment that has been assigned to you.
- Be prepared for your lab by reading the experiment ahead of time and completing any pre-lab activities if there are any.
- For many laboratory experiments, safety eyewear should be worn at all times. Consult your instructor about the need for and kind of safety eyewear, such as goggles, and determine the policy for wearing contact lenses in the laboratory.
- Dress appropriately for lab, with clothing that covers your torso and legs and shoes that cover your entire foot. Tie back loose hair and avoid wearing baggy sleeves or dangling jewelry. Your instructor will inform you if you are to wear a lab coat or lab apron.
- For some experiments you may be required to wear gloves. Be sure to remove the gloves when leaving the lab even if you are planning to return right away. In addition, always wash your hands before leaving the lab.
- Eating, drinking, applying makeup, etc., are forbidden in the laboratory.

- Pipetting by mouth is never allowed in the laboratory.
- Dispose of all chemical waste or other waste that is produced during the lab in an appropriate manner designated by your instructor.
- Be sure you know the location of all safety equipment such as the safety shower, eyewash fountains, fire extinguishers, and emergency telephones. Report any accident, breakage of glassware, or spills to your instructor immediately.
- It is important to know how to evacuate from the laboratory in the event of a fire or other emergency.

These general safety guidelines are designed to help keep your laboratory environment as safe as it can be. It is important that you and the other students in the lab act responsibly and be diligent in following all safety rules outlined here and given to you by your instructor. Please sign below to acknowledge that you have read the guidelines and agree to follow them and any other directives given to you by your instructor.

______________________________ Student Signature

________________ Date

Laboratory Safety

Laboratory safety should be the number one priority for you, your fellow students, and your instructors. The guidelines seen below are general laboratory safety practices that can apply to most laboratory settings. Your instructor will advise you of any specific precautions or hazards you may deal with in a particular experiment. Do not hesitate to ask your instructor about any questions you might have about an experiment or any of the safety precautions outlined here or given to you by your instructor.

- Never work alone in the lab without supervision and work only on an experiment that has been assigned to you.
- Be prepared for your lab by reading the experiment ahead of time and completing any pre-lab activities if there are any.
- For many laboratory experiments, safety eyewear should be worn at all times. Consult your instructor about the need for and kind of safety eyewear, such as goggles, and determine the policy for wearing contact lenses in the laboratory.
- Dress appropriately for lab, with clothing that covers your torso and legs and shoes that cover your entire foot. Tie back loose hair and avoid wearing baggy sleeves or dangling jewelry. Your instructor will inform you if you are to wear a lab coat or lab apron.
- For some experiments you may be required to wear gloves. Be sure to remove the gloves when leaving the lab even if you are planning to return right away. In addition, always wash your hands before leaving the lab.
- Eating, drinking, applying makeup, etc., are forbidden in the laboratory.

- Pipetting by mouth is never allowed in the laboratory.
- Dispose of all chemical waste or other waste that is produced during the lab in an appropriate manner designated by your instructor.
- Be sure you know the location of all safety equipment such as the safety shower, eyewash fountains, fire extinguishers, and emergency telephones. Report any accident, breakage of glassware, or spills to your instructor immediately.
- It is important to know how to evacuate from the laboratory in the event of a fire or other emergency.

These general safety guidelines are designed to help keep your laboratory environment as safe as it can be. It is important that you and the other students in the lab act responsibly and be diligent in following all safety rules outlined here and given to you by your instructor. Please sign below to acknowledge that you have read the guidelines and agree to follow them and any other directives given to you by your instructor.

______________________________ _______________

Student Signature Date

Exp. No.	Experiment/Subject		Date	
Name		Lab Partner	Locker/ Desk No.	Course & Section No.

Signature	Date	Witness/TA	Date

Exp. No.	Experiment/Subject	Date

Name	Lab Partner	Locker/ Desk No.	Course & Section No.

Signature	Date	Witness/TA	Date

Note: Place fold-over back cover under copy sheet before writing

Exp. No.	Experiment/Subject		Date	
Name		Lab Partner	Locker/ Desk No.	Course & Section No.

Signature	Date	Witness/TA	Date

Exp. No.	Experiment/Subject		Date	
Name		Lab Partner	Locker/ Desk No.	Course & Section No.

Signature	Date	Witness/TA	Date

Note: Place fold-over back cover under copy sheet before writing

Exp. No.	Experiment/Subject		Date	
Name		Lab Partner	Locker/ Desk No.	Course & Section No.

Signature	Date	Witness/TA	Date

Note: Place fold-over back cover under copy sheet before writing

Exp. No.	Experiment/Subject		Date	
Name		Lab Partner	Locker/ Desk No.	Course & Section No.

Signature	Date	Witness/TA	Date

Note: Place fold-over back cover under copy sheet before writing

Exp. No.	Experiment/Subject	Date	
Name	Lab Partner	Locker/ Desk No.	Course & Section No.

Signature	Date	Witness/TA	Date

Note: Place fold-over back cover under copy sheet before writing

Exp. No.	Experiment/Subject		Date	
Name		Lab Partner	Locker/ Desk No.	Course & Section No.

Signature	Date	Witness/TA	Date

Note: Place fold-over back cover under copy sheet before writing

Exp. No.	Experiment/Subject		Date	
Name		Lab Partner	Locker/ Desk No.	Course & Section No.

Signature	Date	Witness/TA	Date

Note: Place fold-over back cover under copy sheet before writing

Exp. No.	Experiment/Subject	Date	
Name	Lab Partner	Locker/ Desk No.	Course & Section No.

Signature	Date	Witness/TA	Date

Exp. No.	Experiment/Subject	Date	
Name	Lab Partner	Locker/ Desk No.	Course & Section No.

Signature	Date	Witness/TA	Date

Note: Place fold-over back cover under copy sheet before writing

Exp. No.	Experiment/Subject		Date	
Name		Lab Partner	Locker/ Desk No.	Course & Section No.

Signature	Date	Witness/TA	Date

Exp. No.	Experiment/Subject		Date	
Name	Lab Partner		Locker/ Desk No.	Course & Section No.

Signature	Date	Witness/TA	Date

Note: Place fold-over back cover under copy sheet before writing

Exp. No.	Experiment/Subject		Date	
Name		Lab Partner	Locker/ Desk No.	Course & Section No.

Signature	Date	Witness/TA	Date

Note: Place fold-over back cover under copy sheet before writing

Exp. No.	Experiment/Subject	Date	
Name	Lab Partner	Locker/ Desk No.	Course & Section No.

Signature	Date	Witness/TA	Date

Note: Place fold-over back cover under copy sheet before writing

Exp. No.	Experiment/Subject		Date	
Name		Lab Partner	Locker/ Desk No.	Course & Section No.

Signature	Date	Witness/TA	Date

Note: Place fold-over back cover under copy sheet before writing

Exp. No.	Experiment/Subject	Date	
Name	Lab Partner	Locker/ Desk No.	Course & Section No.

Signature	Date	Witness/TA	Date

Exp. No.	Experiment/Subject		Date	
Name		Lab Partner	Locker/ Desk No.	Course & Section No.

COPY

Signature	Date	Witness/TA	Date

Note: Place fold-over back cover under copy sheet before writing

Exp. No.	Experiment/Subject		Date	
Name		Lab Partner	Locker/ Desk No.	Course & Section No.

Signature	Date	Witness/TA	Date

Note: Place fold-over back cover under copy sheet before writing

Exp. No.	Experiment/Subject		Date	
Name		Lab Partner	Locker/ Desk No.	Course & Section No.

Signature	Date	Witness/TA	Date

Exp. No.	Experiment/Subject	Date	
Name	Lab Partner	Locker/ Desk No.	Course & Section No.

Signature	Date	Witness/TA	Date

Note: Place fold-over back cover under copy sheet before writing

Exp. No.	Experiment/Subject		Date	
Name		Lab Partner	Locker/ Desk No.	Course & Section No.

Signature	Date	Witness/TA	Date

Note: Place fold-over back cover under copy sheet before writing

Exp. No.	Experiment/Subject	Date	
Name	Lab Partner	Locker/ Desk No.	Course & Section No.

Signature	Date	Witness/TA	Date

Note: Place fold-over back cover under copy sheet before writing

Exp. No.	Experiment/Subject		Date	
Name		Lab Partner	Locker/ Desk No.	Course & Section No.

Signature	Date	Witness/TA	Date

Note: Place fold-over back cover under copy sheet before writing

Exp. No.	Experiment/Subject		Date	
Name		Lab Partner	Locker/ Desk No.	Course & Section No.

Signature	Date	Witness/TA	Date

Exp. No.	Experiment/Subject		Date	
Name		Lab Partner	Locker/ Desk No.	Course & Section No.

Signature	Date	Witness/TA	Date

Note: Place fold-over back cover under copy sheet before writing

Exp. No.	Experiment/Subject		Date	
Name		Lab Partner	Locker/ Desk No.	Course & Section No.

Signature	Date	Witness/TA	Date

Exp. No.	Experiment/Subject		Date	
Name		Lab Partner	Locker/ Desk No.	Course & Section No.

Signature	Date	Witness/TA	Date

Exp. No.	Experiment/Subject	Date	
Name	Lab Partner	Locker/ Desk No.	Course & Section No.

Signature	Date	Witness/TA	Date

Exp. No.	Experiment/Subject		Date	
Name		Lab Partner	Locker/ Desk No.	Course & Section No.

Signature	Date	Witness/TA	Date

Note: Place fold-over back cover under copy sheet before writing

Exp. No.	Experiment/Subject		Date	
Name	Lab Partner		Locker/ Desk No.	Course & Section No.

Signature	Date	Witness/TA	Date

Exp. No.	Experiment/Subject		Date	
Name		Lab Partner	Locker/ Desk No.	Course & Section No.

Signature	Date	Witness/TA	Date

Note: Place fold-over back cover under copy sheet before writing

Exp. No.	Experiment/Subject	Date	
Name	Lab Partner	Locker/ Desk No.	Course & Section No.

Signature	Date	Witness/TA	Date

Note: Place fold-over back cover under copy sheet before writing

Exp. No.	Experiment/Subject		Date	
Name		Lab Partner	Locker/ Desk No.	Course & Section No.

Signature	Date	Witness/TA	Date

Exp. No.	Experiment/Subject		Date	
Name		Lab Partner	Locker/ Desk No.	Course & Section No.

Signature	Date	Witness/TA	Date

Exp. No.	Experiment/Subject		Date	
Name		Lab Partner	Locker/ Desk No.	Course & Section No.

Signature	Date	Witness/TA	Date

Exp. No.	Experiment/Subject		Date	
Name		Lab Partner	Locker/ Desk No.	Course & Section No.

Signature	Date	Witness/TA	Date

Note: Place fold-over back cover under copy sheet before writing

Exp. No.	Experiment/Subject		Date	
Name	Lab Partner		Locker/ Desk No.	Course & Section No.

Signature	Date	Witness/TA	Date

Exp. No.	Experiment/Subject		Date	
Name		Lab Partner	Locker/ Desk No.	Course & Section No.

Signature	Date	Witness/TA	Date

Note: Place fold-over back cover under copy sheet before writing

Exp. No.	Experiment/Subject	Date	
Name	Lab Partner	Locker/ Desk No.	Course & Section No.

Signature	Date	Witness/TA	Date

Note: Place fold-over back cover under copy sheet before writing

Exp. No.	Experiment/Subject		Date	
Name		Lab Partner	Locker/ Desk No.	Course & Section No.

Signature	Date	Witness/TA	Date

Note: Place fold-over back cover under copy sheet before writing

Exp. No.	Experiment/Subject		Date	
Name		Lab Partner	Locker/ Desk No.	Course & Section No.

Signature	Date	Witness/TA	Date

Exp. No.	Experiment/Subject	Date

Name	Lab Partner	Locker/ Desk No.	Course & Section No.

Signature	Date	Witness/TA	Date

Note: Place fold-over back cover under copy sheet before writing

Exp. No.	Experiment/Subject	Date	
Name	Lab Partner	Locker/ Desk No.	Course & Section No.

COPY

Signature	Date	Witness/TA	Date

Note: Place fold-over back cover under copy sheet before writing

Exp. No.	Experiment/Subject		Date	
Name		Lab Partner	Locker/ Desk No.	Course & Section No.

Signature	Date	Witness/TA	Date

Note: Place fold-over back cover under copy sheet before writing

Exp. No.	Experiment/Subject		Date	
Name		Lab Partner	Locker/ Desk No.	Course & Section No.

Signature	Date	Witness/TA	Date

Note: Place fold-over back cover under copy sheet before writing

Exp. No.	Experiment/Subject	Date	
Name	Lab Partner	Locker/ Desk No.	Course & Section No.

Signature	Date	Witness/TA	Date

Note: Place fold-over back cover under copy sheet before writing

Exp. No.	Experiment/Subject		Date	
Name		Lab Partner	Locker/ Desk No.	Course & Section No.

Signature	Date	Witness/TA	Date

Exp. No.	Experiment/Subject		Date	
Name		Lab Partner	Locker/ Desk No.	Course & Section No.

Signature	Date	Witness/TA	Date

Note: Place fold-over back cover under copy sheet before writing

Exp. No.	Experiment/Subject		Date	
Name		Lab Partner	Locker/ Desk No.	Course & Section No.

Signature	Date	Witness/TA	Date

Note: Place fold-over back cover under copy sheet before writing

Exp. No.	Experiment/Subject		Date	
Name		Lab Partner	Locker/ Desk No.	Course & Section No.

Signature	Date	Witness/TA	Date

Exp. No.	Experiment/Subject		Date	
Name		Lab Partner	Locker/ Desk No.	Course & Section No.

Signature	Date	Witness/TA	Date

Note: Place fold-over back cover under copy sheet before writing

Exp. No.	Experiment/Subject		Date	
Name		Lab Partner	Locker/ Desk No.	Course & Section No.

Signature	Date	Witness/TA	Date

Note: Place fold-over back cover under copy sheet before writing

Exp. No.	Experiment/Subject		Date	
Name		Lab Partner	Locker/ Desk No.	Course & Section No.

Signature	Date	Witness/TA	Date

Exp. No.	Experiment/Subject		Date	
Name	Lab Partner		Locker/ Desk No.	Course & Section No.

Signature	Date	Witness/TA	Date

Note: Place fold-over back cover under copy sheet before writing

Exp. No.	Experiment/Subject	Date	
Name	Lab Partner	Locker/ Desk No.	Course & Section No.

Signature	Date	Witness/TA	Date

Note: Place fold-over back cover under copy sheet before writing

Exp. No.	Experiment/Subject		Date	
Name		Lab Partner	Locker/ Desk No.	Course & Section No.

Signature	Date	Witness/TA	Date

Note: Place fold-over back cover under copy sheet before writing

Exp. No.	Experiment/Subject	Date	
Name	Lab Partner	Locker/ Desk No.	Course & Section No.

Signature	Date	Witness/TA	Date

Exp. No.	Experiment/Subject		Date	
Name		Lab Partner	Locker/ Desk No.	Course & Section No.

Signature	Date	Witness/TA	Date

Note: Place fold-over back cover under copy sheet before writing

Exp. No.	Experiment/Subject		Date	
Name		Lab Partner	Locker/ Desk No.	Course & Section No.

COPY

Signature	Date	Witness/TA	Date

Note: Place fold-over back cover under copy sheet before writing

Exp. No.	Experiment/Subject	Date	
Name	Lab Partner	Locker/ Desk No.	Course & Section No.

Signature	Date	Witness/TA	Date

Exp. No.	Experiment/Subject	Date	
Name	Lab Partner	Locker/ Desk No.	Course & Section No.

Signature	Date	Witness/TA	Date

Exp. No.	Experiment/Subject		Date	
Name		Lab Partner	Locker/ Desk No.	Course & Section No.

Signature	Date	Witness/TA	Date

Exp. No.	Experiment/Subject		Date	
Name		Lab Partner	Locker/ Desk No.	Course & Section No.

Signature	Date	Witness/TA	Date

Note: Place fold-over back cover under copy sheet before writing

Exp. No.	Experiment/Subject	Date	
Name	Lab Partner	Locker/ Desk No.	Course & Section No.

Signature	Date	Witness/TA	Date

Note: Place fold-over back cover under copy sheet before writing

Exp. No.	Experiment/Subject	Date	
Name	Lab Partner	Locker/ Desk No.	Course & Section No.

COPY

Signature	Date	Witness/TA	Date

Note: Place fold-over back cover under copy sheet before writing

Exp. No.	Experiment/Subject		Date	
Name		Lab Partner	Locker/ Desk No.	Course & Section No.

Signature	Date	Witness/TA	Date

Note: Place fold-over back cover under copy sheet before writing

Exp. No.	Experiment/Subject		Date	
Name		Lab Partner	Locker/ Desk No.	Course & Section No.

Signature	Date	Witness/TA	Date

Note: Place fold-over back cover under copy sheet before writing

Exp. No.	Experiment/Subject	Date	
Name	Lab Partner	Locker/ Desk No.	Course & Section No.

Signature	Date	Witness/TA	Date

Note: Place fold-over back cover under copy sheet before writing

Exp. No.	Experiment/Subject		Date	
Name		Lab Partner	Locker/ Desk No.	Course & Section No.

Signature	Date	Witness/TA	Date

Note: Place fold-over back cover under copy sheet before writing

Exp. No.	Experiment/Subject		Date	
Name		Lab Partner	Locker/ Desk No.	Course & Section No.

Signature	Date	Witness/TA	Date

Note: Place fold-over back cover under copy sheet before writing

Exp. No.	Experiment/Subject		Date	
Name		Lab Partner	Locker/ Desk No.	Course & Section No.

Signature	Date	Witness/TA	Date

Note: Place fold-over back cover under copy sheet before writing

Exp. No.	Experiment/Subject		Date	
Name		Lab Partner	Locker/ Desk No.	Course & Section No.

Signature	Date	Witness/TA	Date

Note: Place fold-over back cover under copy sheet before writing

Exp. No.	Experiment/Subject		Date	
Name		Lab Partner	Locker/ Desk No.	Course & Section No.

Signature	Date	Witness/TA	Date

Note: Place fold-over back cover under copy sheet before writing

Exp. No.	Experiment/Subject	Date	
Name	Lab Partner	Locker/ Desk No.	Course & Section No.

Signature	Date	Witness/TA	Date

Note: Place fold-over back cover under copy sheet before writing

Exp. No.	Experiment/Subject		Date	
Name		Lab Partner	Locker/ Desk No.	Course & Section No.

Signature	Date	Witness/TA	Date

Note: Place fold-over back cover under copy sheet before writing

Exp. No.	Experiment/Subject		Date	
Name		Lab Partner	Locker/ Desk No.	Course & Section No.

Signature	Date	Witness/TA	Date

Exp. No.	Experiment/Subject		Date	
Name		Lab Partner	Locker/ Desk No.	Course & Section No.

Signature	Date	Witness/TA	Date

Exp. No.	Experiment/Subject		Date	
Name		Lab Partner	Locker/ Desk No.	Course & Section No.

Signature	Date	Witness/TA	Date

Exp. No.	Experiment/Subject		Date	
Name		Lab Partner	Locker/ Desk No.	Course & Section No.

Signature	Date	Witness/TA	Date

Note: Place fold-over back cover under copy sheet before writing

Exp. No.	Experiment/Subject	Date	
Name	Lab Partner	Locker/ Desk No.	Course & Section No.

Signature	Date	Witness/TA	Date

Note: Place fold-over back cover under copy sheet before writing

Exp. No.	Experiment/Subject		Date	
Name		Lab Partner	Locker/ Desk No.	Course & Section No.

Signature	Date	Witness/TA	Date

Exp. No.	Experiment/Subject	Date	
Name	Lab Partner	Locker/ Desk No.	Course & Section No.

Signature	Date	Witness/TA	Date

Note: Place fold-over back cover under copy sheet before writing

Exp. No.	Experiment/Subject		Date	
Name		Lab Partner	Locker/ Desk No.	Course & Section No.

COPY

Signature	Date	Witness/TA	Date

Note: Place fold-over back cover under copy sheet before writing

Exp. No.	Experiment/Subject	Date

Name	Lab Partner	Locker/ Desk No.	Course & Section No.

Signature	Date	Witness/TA	Date

Exp. No.	Experiment/Subject		Date	
Name		Lab Partner	Locker/ Desk No.	Course & Section No.

Signature	Date	Witness/TA	Date

Exp. No.	Experiment/Subject		Date	
Name		Lab Partner	Locker/ Desk No.	Course & Section No.

Signature	Date	Witness/TA	Date

Note: Place fold-over back cover under copy sheet before writing

Exp. No.	Experiment/Subject		Date	
Name		Lab Partner	Locker/ Desk No.	Course & Section No.

Signature	Date	Witness/TA	Date

Exp. No.	Experiment/Subject	Date

Name	Lab Partner	Locker/ Desk No.	Course & Section No.

Signature	Date	Witness/TA	Date

Note: Place fold-over back cover under copy sheet before writing

Exp. No.	Experiment/Subject		Date	
Name		Lab Partner	Locker/ Desk No.	Course & Section No.

Signature	Date	Witness/TA	Date

Note: Place fold-over back cover under copy sheet before writing

Exp. No.	Experiment/Subject		Date	
Name		Lab Partner	Locker/ Desk No.	Course & Section No.

Signature	Date	Witness/TA	Date

Exp. No.	Experiment/Subject		Date	
Name		Lab Partner	Locker/ Desk No.	Course & Section No.

Signature	Date	Witness/TA	Date

Exp. No.	Experiment/Subject		Date	
Name		Lab Partner	Locker/ Desk No.	Course & Section No.

Signature	Date	Witness/TA	Date

Note: Place fold-over back cover under copy sheet before writing

Exp. No.	Experiment/Subject		Date	
Name		Lab Partner	Locker/ Desk No.	Course & Section No.

Signature	Date	Witness/TA	Date

Note: Place fold-over back cover under copy sheet before writing

Exp. No.	Experiment/Subject	Date	
Name	Lab Partner	Locker/ Desk No.	Course & Section No.

Signature	Date	Witness/TA	Date

Note: Place fold-over back cover under copy sheet before writing

Exp. No.	Experiment/Subject		Date	
Name		Lab Partner	Locker/ Desk No.	Course & Section No.

Signature	Date	Witness/TA	Date

Exp. No.	Experiment/Subject		Date	
Name		Lab Partner	Locker/ Desk No.	Course & Section No.

Signature	Date	Witness/TA	Date

Exp. No.	Experiment/Subject		Date	
Name	Lab Partner		Locker/ Desk No.	Course & Section No.

Signature	Date	Witness/TA	Date

Note: Place fold-over back cover under copy sheet before writing

Exp. No.	Experiment/Subject	Date	
Name	Lab Partner	Locker/ Desk No.	Course & Section No.

Signature	Date	Witness/TA	Date

Note: Place fold-over back cover under copy sheet before writing

Exp. No.	Experiment/Subject		Date	
Name		Lab Partner	Locker/ Desk No.	Course & Section No.

Signature	Date	Witness/TA	Date

Note: Place fold-over back cover under copy sheet before writing